设计有约 2（上册）

inSIDE deSign

香港方黄建筑师事务所私邸空间设计系列专集 方峻 著

华中科技大学出版社
http://www.hustp.com
中国·武汉

TFong, a multi and trans-industry designer
in Hong Kong, China. studied the bachelor/postgraduate/doctor
programs and degrees on philosophy, construction engineering and
design management in Inter American University,
Politecnico di Milano, Huaqiao University, and Hong Kong
Polytechnic University. He is a Professional Associations of Hong Kong
Interior Design Association (PM00408),
Member of International Federation of Interior Architects/ Interior Designers (0281),
Institute of Interior Design of Architectural Society of China (8030),
Lighting designer of register of China.
The works not only was awarded the golden award
at Indoor Design Contest for the 1st International Building Landscape,
also was awarded by awards and special honors
at indoor design contests home and abroad.
Be more have been inserted by American *Indoor Design* and
various well-known magazines.
Here are published and highly acclaimed industry of
personal album design have been
Inspiration Design,
Inside Design I, Inside Design II...Inside Design V,
Mangement System and Application for Installation Art Project etc.

方峻（TFong）

建筑空间与多元跨界的中国香港设计师。

先后在美国美联大学、意大利米兰理工学院、香港理工大学、国立华侨大学

接受哲学、建筑设计、设计管理的学士／硕士／博士等教育，

同时也是香港室内设计协会专业会员、国际室内建筑师设计联盟会员、中国建筑学会室内设计分会会员、中国注册照明设计师。

其作品不但荣获首届国际建筑景观室内设计大奖赛金奖，

还获得过多项国内外室内设计的奖项与荣誉；更被美国《室内设计》和

各类知名专业杂志数次刊载。相继出版且备受业界好评的

个人设计专辑分别有《"悟"设计》《设计有约 1》《设计有约 2》……《设计有约 5》

Contents
目录

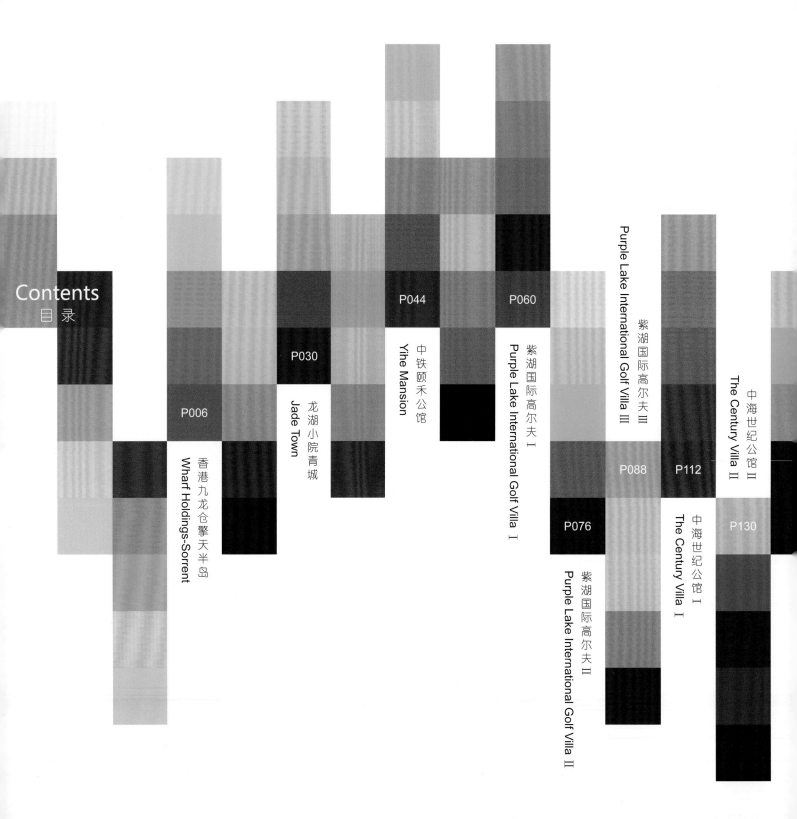

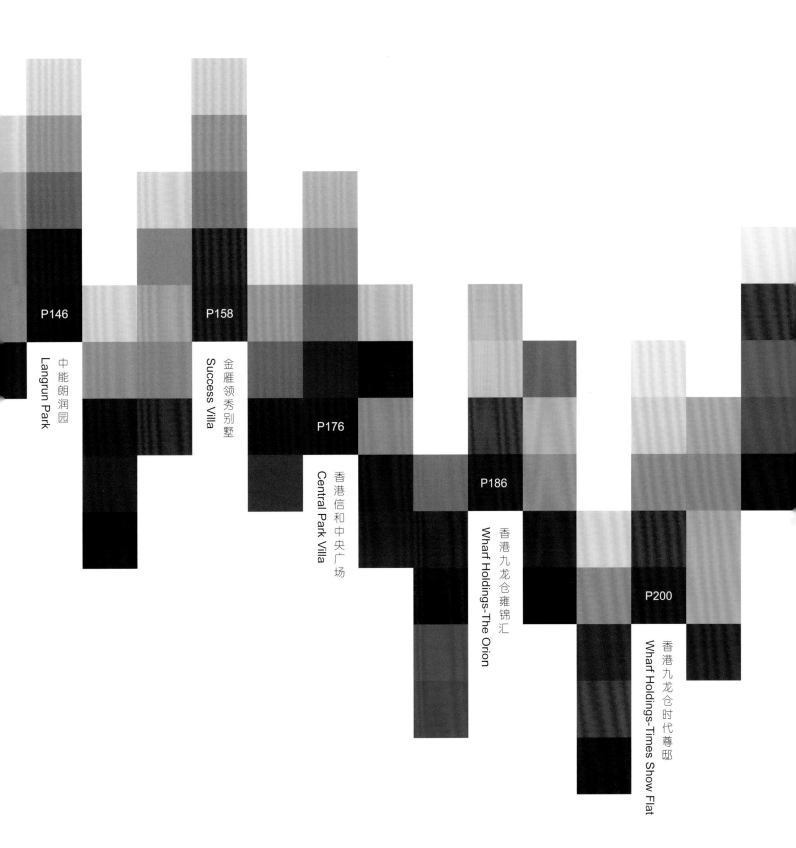

Wharf Holdings-Sorrent

香港九龙仓擎天半岛

The refreshing embellishment of vivid yellow accentuates the space originally set in a cold color tone of white and grey with vibrancy and a splash of life. The yellowish hue amiably lightens up the serene dignity and transforms it into shimmers of elegance.

棣棠花鲜嫩的黄色作为白色、灰色等冷色调的点缀，将空间的色彩感一下子渲染出来，生动而活泼。在黄色的点缀下，静谧而有些许庄重的气氛略显轻松，又增添了几分高雅。

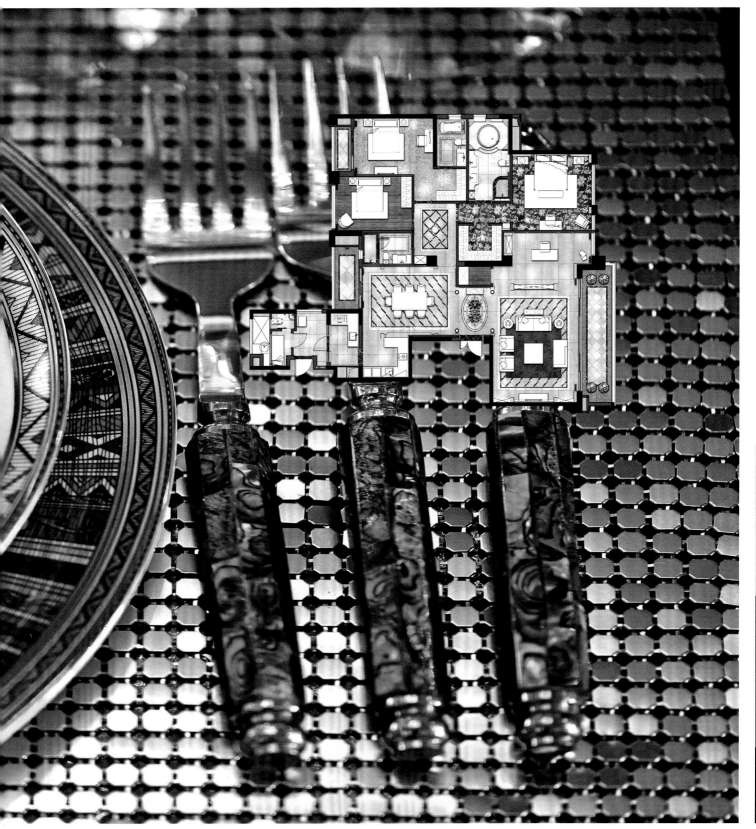

Jade Town

龙湖小院青城

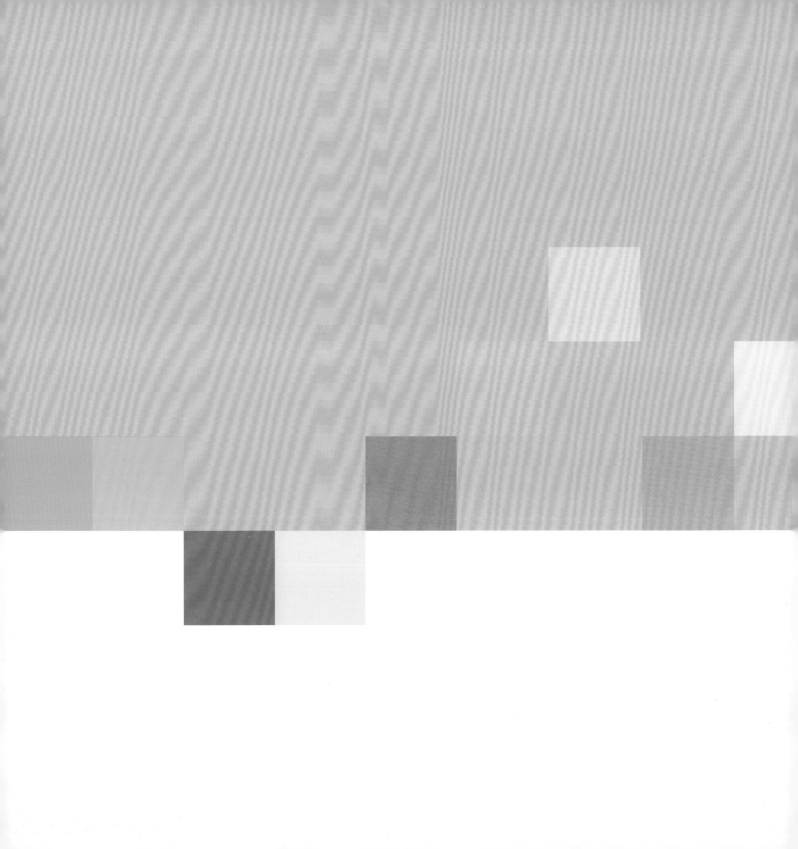

Yihe Mansion

中铁颐禾公馆

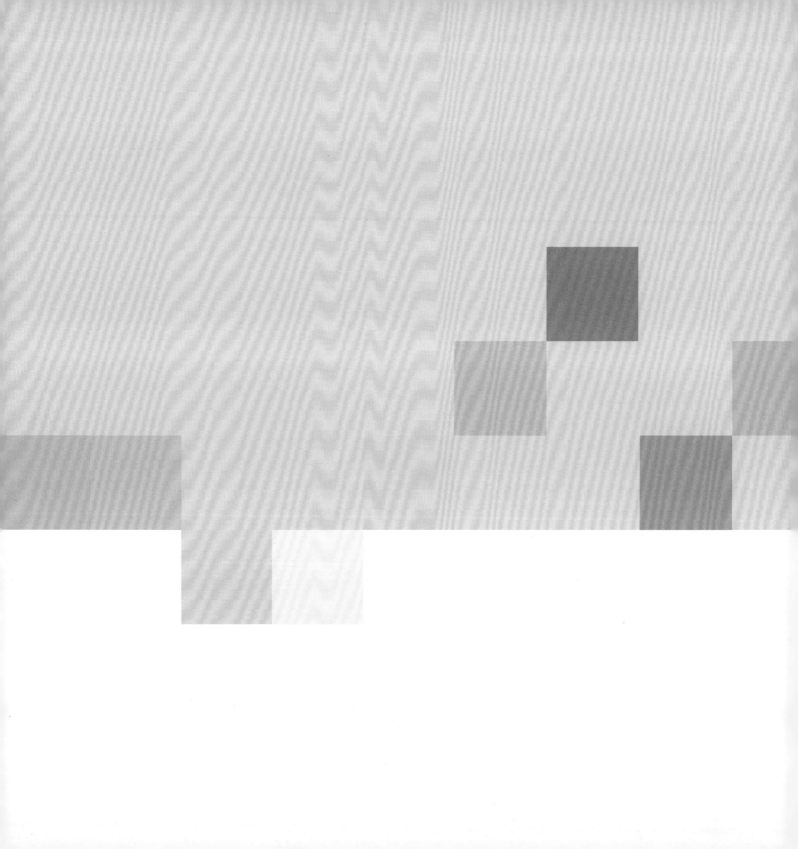

Purple Lake International Golf Villa Ⅰ
紫湖国际高尔夫Ⅰ

abundant and beautiful: grain yellow. Nothing can better present a sensation of a bumper harvest than grain gold. The combination of abundance, glory and beauty brings an intense feeling of happiness. Also known as sun-orange, this color is associated with the harvest season because the orange is said to be a symbol of fertility since the Rome times. In practice, it embodies wealth and happiness, which makes it a suitable color for family style.

没有什么色彩能比稻谷金更能呈现丰收之感了。丰富、光辉和美的结合，使这种色彩给人带来强烈的幸福感。这种色亦为"SUN-ORANGE"，据说在欧洲，橙子的果实从罗马时代开始就被当作多产的象征，因此从这个色彩就很容易令人联想到收获的季节。在运用中，它象征着富贵和幸福，也是适合家庭的色彩。

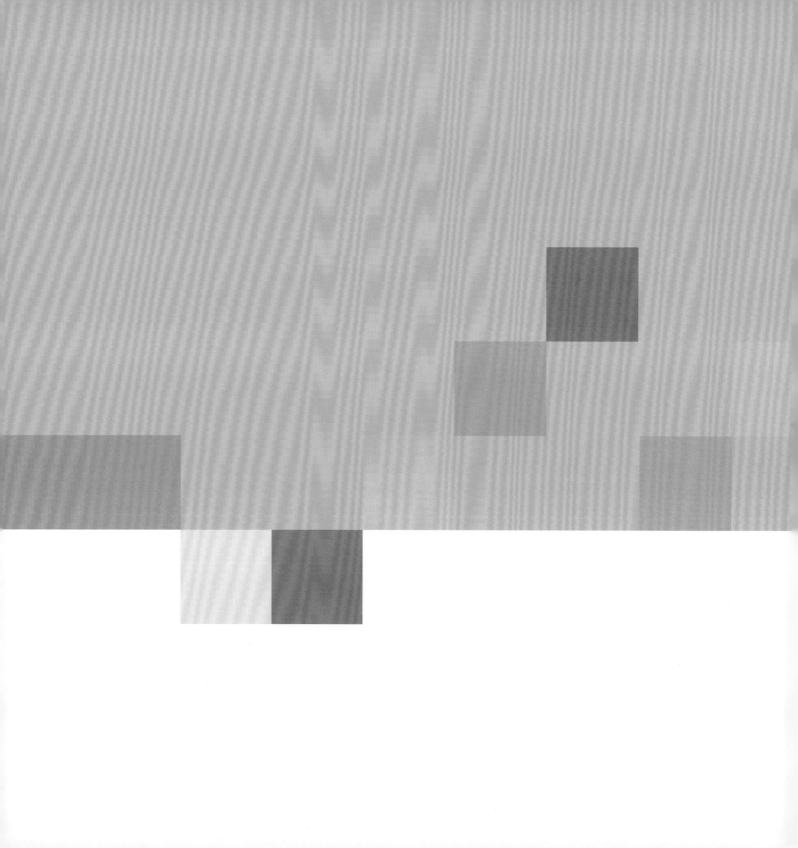

Purple Lake International Golf Villa II

紫湖国际高尔夫 II

luxurious and abundant.pot marigold The name originates from the Calendula officinalis of Southern Europe, and is a traditional color in Britain and North America. As a potent and warm yellow, it reminds one of gold, for which the pot marigold symbolizes abundance, glory and beauty. The incorporation of the disposition of red strengthens its impact and sense of happiness.

金盏菊，奢华富丽。这个色彩的名称源自原产于南欧的金盏花，在英美是很受欢迎也是很传统的色彩，它强而温暖的黄色，能让人联想到黄金，所以金盏花象征着丰富、光辉和美丽。因为融合了红色的个性，所以冲击力强，也更突显了黄色的幸福感。

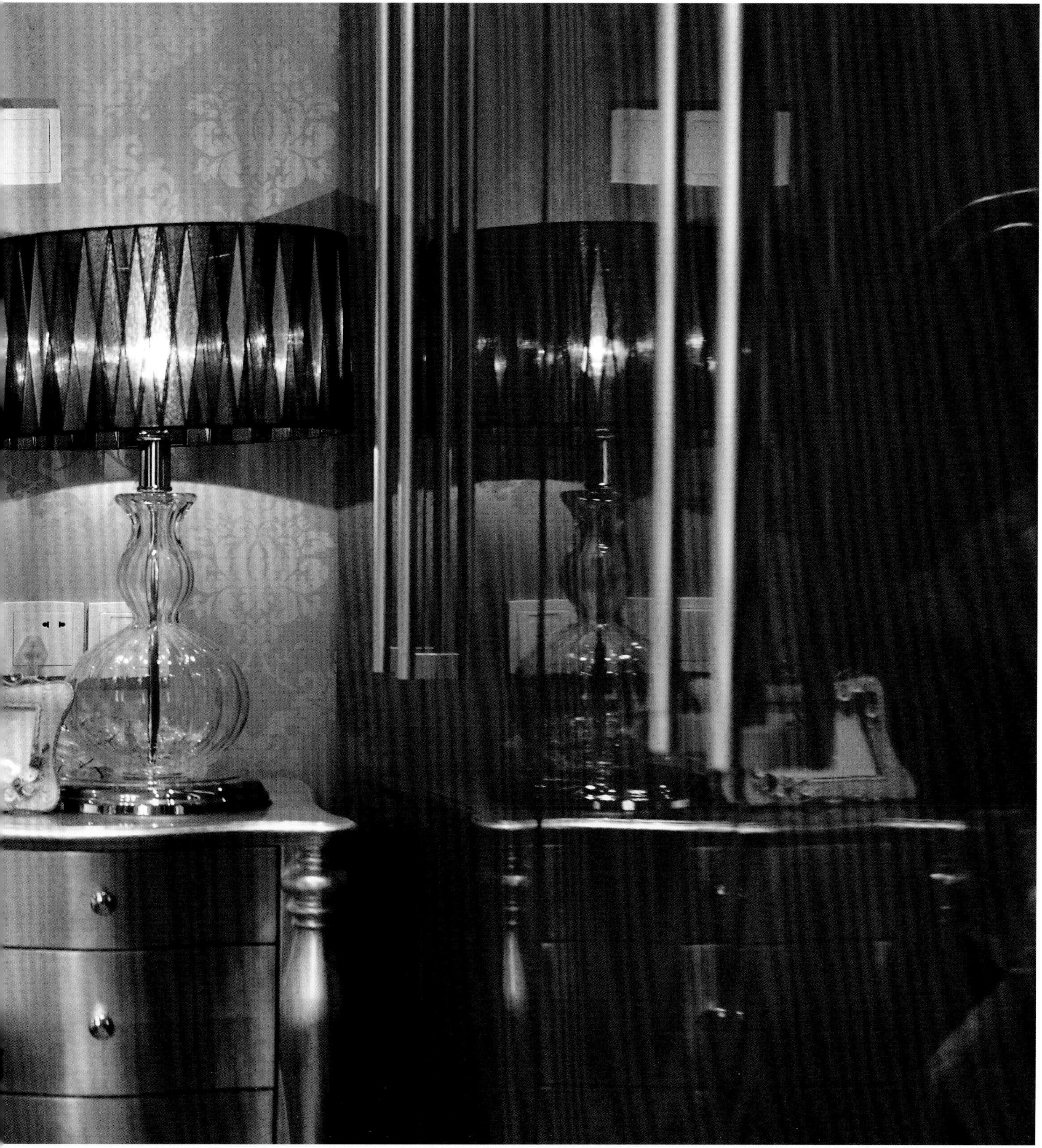

Purple Lake International Golf Villa Ⅲ

紫湖国际高尔夫Ⅲ

Sunshine pours on to tree branches and flowers, and it is so quiet as if you can hear the branches touching the shadow.

Just like the silence before the opening of a good show, what you see after opening the door is another scene with both movement and tranquility in the courtyard.

Flourishingly, the flowers in full bloom are so splendid!

Bright-colored carpet, gorgeous flower arrangement in the vase, well-proportioned picture frames in different shapes on the wall, and similarly bright and flowery curtain rich multicolored decorations are in dizzying quantities for your eyes to see, alive with warm and unrestrained summer love.

阳光倾洒在树枝上与花丛中，似乎静谧得能听到枝头光影碰触的声响，就像一幕好戏开场前的静寂。推门所见，是另一番与庭院动静相宜的场景，繁华似锦。色彩斑斓的地毯，花瓶上绚烂的插花，墙壁上错落有致、形态各异的图框画框，色泽同样明丽香艳的窗帘，花团锦簇，令人目不暇接，洋溢着热烈奔放的夏日情怀。

The Century Villa I

中海世纪公馆 I

Burberry classic card is an easy catch romantic reverie: Scotland plaid, unique texture, elegant design, there is After the pattern of time. Obviously, designers of Burberry a more profound brand interpretation, thus in this project, will be Burberry element clever apply, not dew, classic. The shape Like plus spirit likeness, the spatial pattern of atmospheric fully details with luxury brands Blend, difficult to mask the aristocratic artistically.

Burberry 是一张容易引起人浪漫遐思的经典名片，苏格兰格子，独特的材质，大方优雅的设计，蕴藏在图案后的时代感。显然，设计师对 Burberry 有着较为深层次的品牌解读，因此在此项目中，将 Burberry 的元素巧妙运用，不露痕迹，经典呈现。形似加上神似，空间格局的大气通透与奢华品牌的细节交融，彰显贵族气韵。

The Century Villa II

中海世纪公馆II

Fresh and tender yellow is the important color of the project. This is full of the awaken of spring. The color of fresh yellow, combined with white, gray cool colors, such as the space the color adamantly dyed out, thus the whole ivid and lively. Chinese style Birds and flowers, and the phoenix embroidery figure in the project, the Europe type style is mixed take the oriental element, the light Loose and enchanting, added auspicious and the atmosphere and elegant Emotional appeal.

鲜嫩的黄色是作品的点睛之笔，这种充满春意的新鲜柔美的色彩，再配合白色、灰色等冷色调，将空间的色彩感瞬间渲染出来，从而整使空间显得生动活泼。作品中点缀的中式花鸟及凤凰绣花图，在欧式的风格中混搭了东方元素，轻松妩媚，增添了祥和的气氛及雅致的情调。

Langrun Park

中能朗润园

White hibiscus, quiet and calm. Hibiscus toward dusk fell, but every time fade is to open next more gorgeous, its character, with the spirit of even stronger than before and worship Longed for. White hibiscus is elegant and stick to the model. Hibiscus flowers with white as the background color, brown, blue on collocation Quiet composed such as colour, more foil a each respective characteristics and color, build a harmonious and comfortable atmosphere.

白色木槿花，安静沉稳。木槿花朝开暮落，但每一次凋谢都是为了下一次更绚烂地开放，其矢志弥坚的性格，令人敬慕。白色木槿花绝对是优雅而坚持的典范。空间以木槿花的白色作为背景色，搭配上褐色、蓝色等安静沉稳的色彩，更加衬托出每种色彩各自的特点和美丽，营造出和谐舒适的氛围。

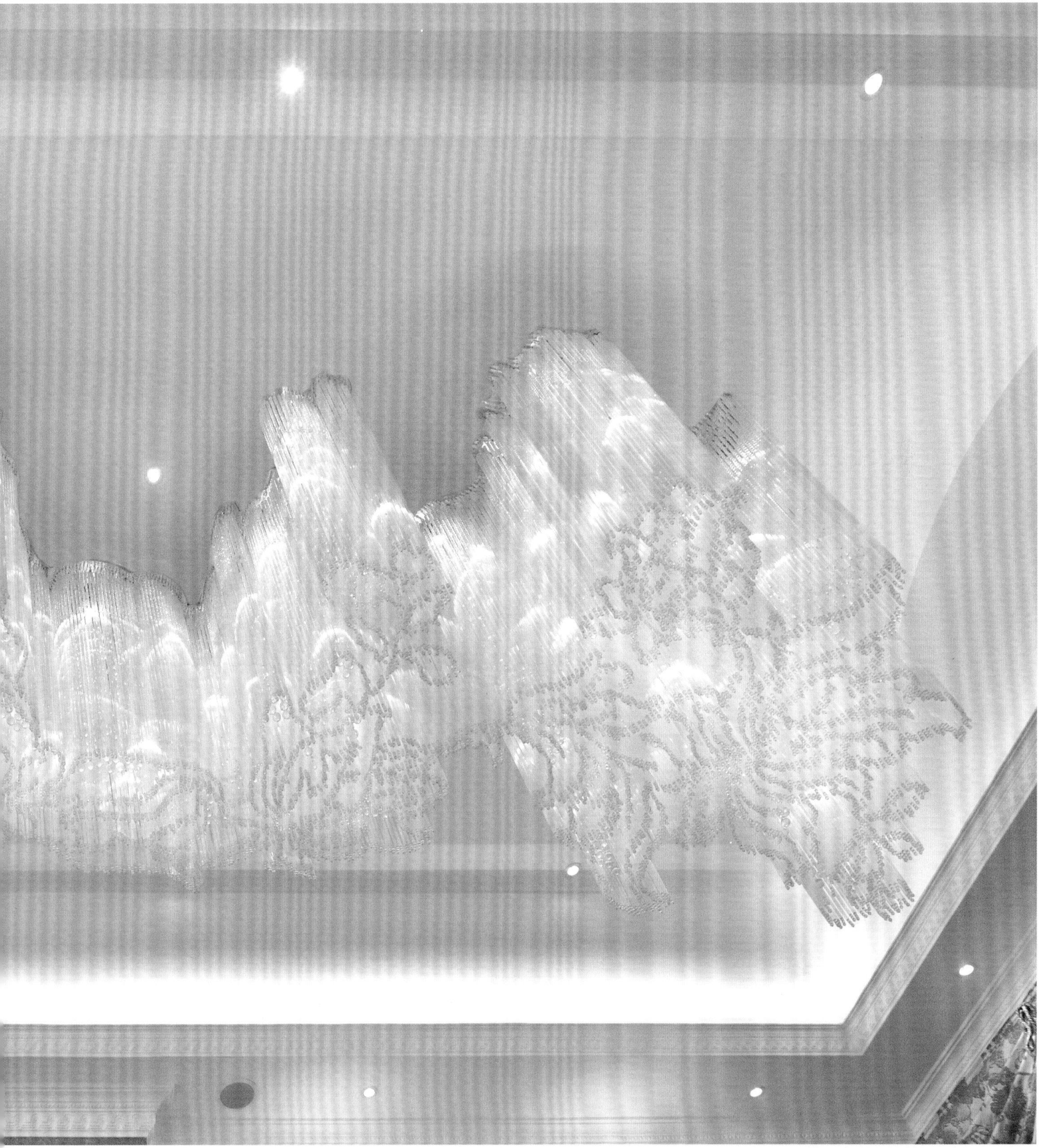

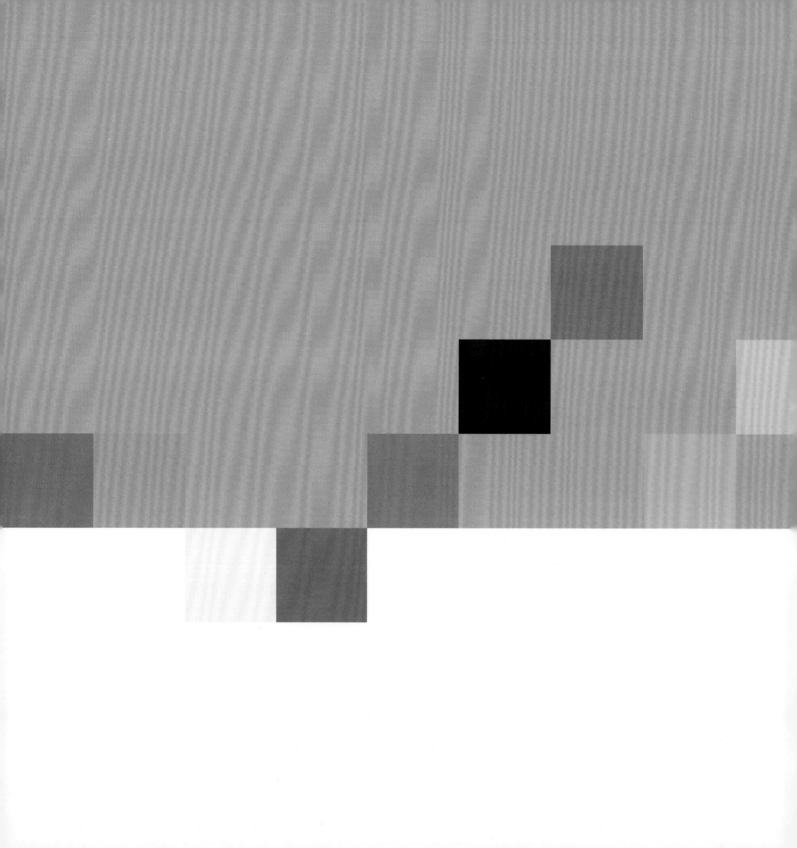

Success Villa

金雁领秀别墅

"Dragon can get large or small, rise or hide; it can summon the cloud and frog while getting large, and hide itself and become invisible when getting small; it can also soar into the universe when rising, and lurk itself into great waves when hiding. It has been deep spring now, dragon will change with the opportunity, as if a person achieves his ambition and becomes invincible around the world. Dragon can be compared to the hero of the world." The villa completed, creatively infused the Chinese element of "dragon" into the classical and European style, so as to reflect the remarkable taste and noble temperament of owners through the extraordinary design.

"龙能大能小，能升能隐；大则兴云吐雾，小则隐介藏形；升则飞腾于宇宙之间，隐则潜伏于波涛之内。方今春深，龙乘时变化，犹人得志而纵横四海，龙之为物，可比世之英雄。"领秀别墅私邸设计，通过巧妙的构思及处理，用中国古典的龙的意象将这只"中国龙"简化变身为"龙"的轻巧符号，体现出业主不凡的品位和高贵的气质。

Central Park Villa

香港信和中央广场

Soft and light: Magnolia. Magnolia is a practically
household name. Its full blossoms are so tender and
fragrant on the trees. Pink purple petals set off
elegant air of love. Pink purple is the absolute color
of grace and softness. Pink may be regarded as
childish, but when matched with potent colors,
it could display a colorful and gorgeous adult world,
and of course, a sweet harmonious and dreamlike
atmosphere.

木兰花，柔和淡雅。这种盛开在小乔木上家喻户晓的花卉，花儿粉妆玉琢，幽香四溢。粉紫色的花瓣
散发着优雅的气息。在体现优雅、柔美时，粉紫色通常是当仁不让的色彩。粉色或许被认为是孩子
气的颜色，但通过与一些强有力的色彩相搭配，也可以展现成人世界的艳丽华美。当然，也可以营
造出甜蜜祥和、宛如梦境般的氛围。

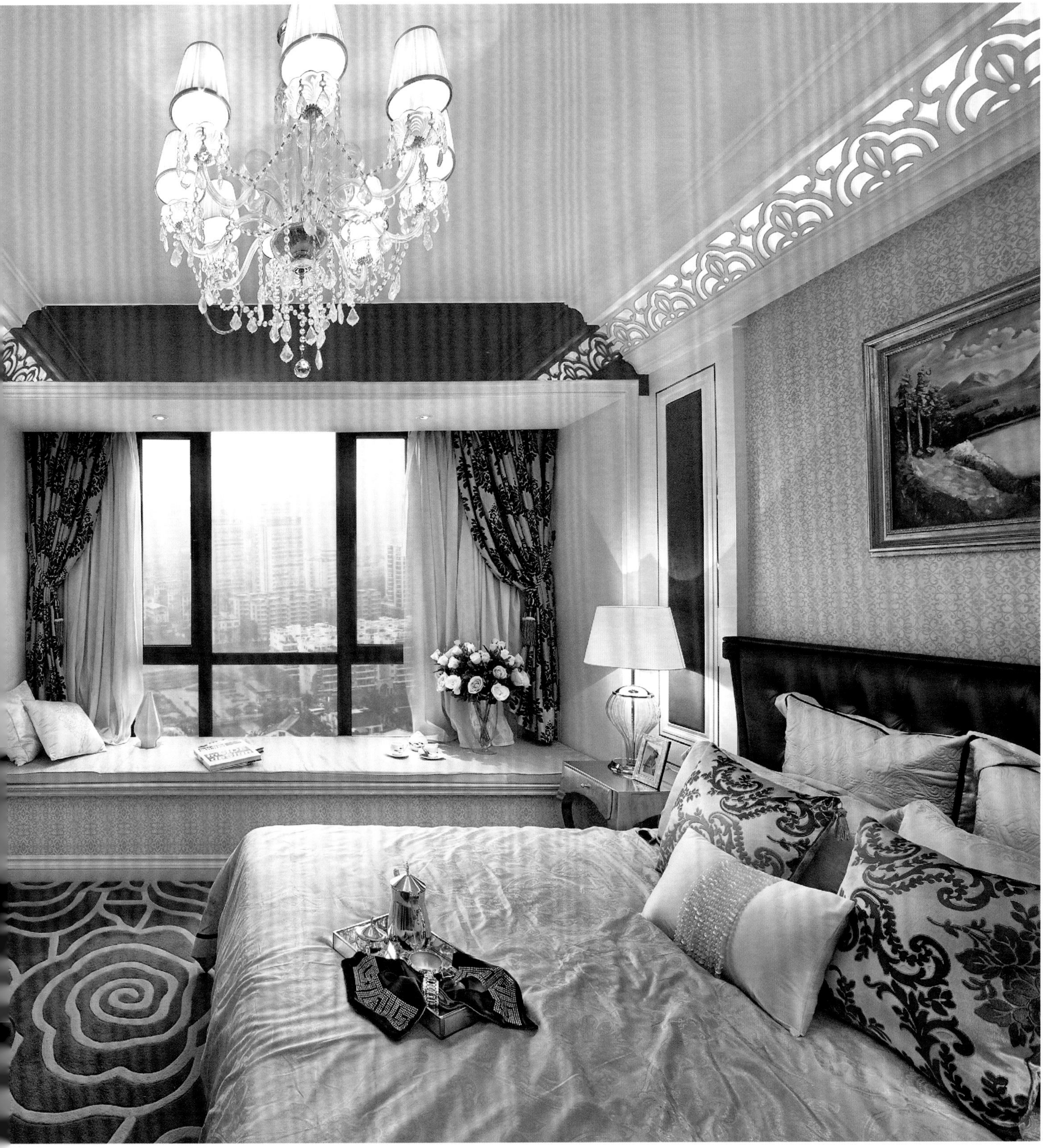

Wharf Holdings-The Orion

香港九龙仓雍锦汇

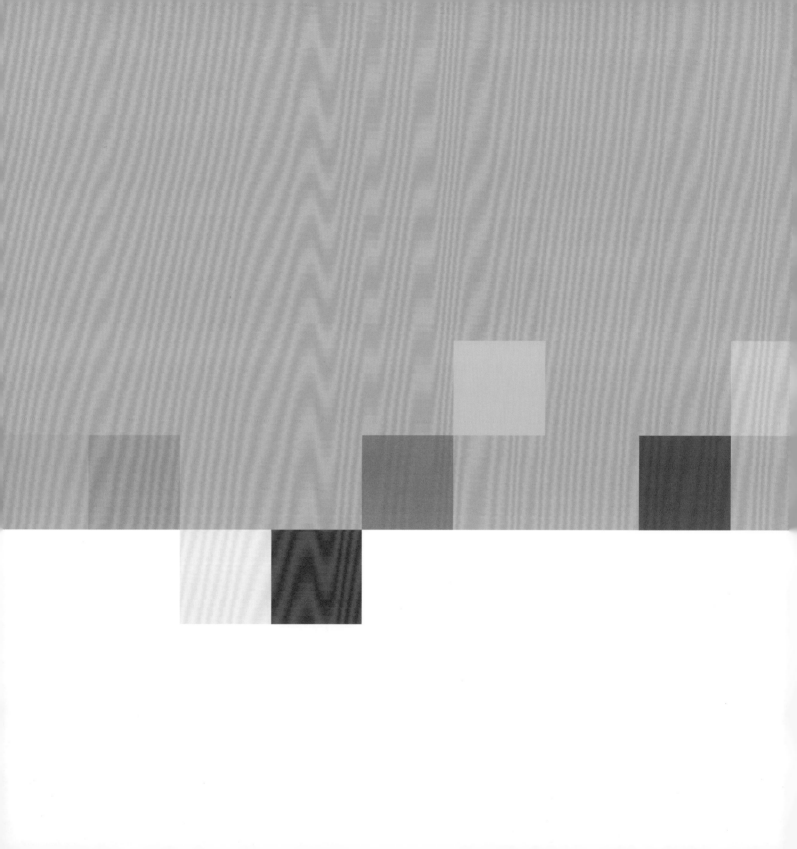

Wharf Holdings-Times Show Flat

香港九龙仓时代尊邸

Gorgeous and brilliant. Barberton daisy originated in South Africa, the Barberton daisy is as a vigorous flower as the sunshine of Africa. Orange chrysanthemums bring red into full play. One cannot but be attracted by its vigor, passion and glory. Still, a cheerful and hilarious scene is created by the match of this bright color and many other colors.

非洲菊，华美亮丽。原产地在南非的非洲菊，如同非洲的阳光一样，是一种能量充沛的花卉。色彩橘红的菊花给人以热情红色的印象。旺盛的生命力和充沛的激情给人以强大的力量感，夺目的华贵美丽，令人不禁深深为其陶醉。而这样明丽的色彩，通过与多种色彩的组合搭配，又营造出欢快、热闹的氛围。

Acknowledgments 鸣谢

本书得以顺利面世，全赖各方的参与与支持，在此由衷感谢"香港方黄建筑师事务所"全体同仁的努力与付出！也衷心感谢以下合作伙伴对我司的大力支持和信任！

九龙仓（中国香港）	信和（中国香港）	新榕建筑置业（中国澳门）
置地集团（中国香港）	南益（中国香港）	新建利建筑置业（中国澳门）
中海地产	龙湖地产	碧桂园地产
华侨城地产	绿地集团	中铁集团
华润地产	珠江地产	中粮地产
远航地产	远大地产	世茂地产
中能置业	大华地产	长航地产
滕王阁地产	光达地产	恒河地产
华人地产	新希望地产	恒丰地产

The cooperation customers cover both well-known Hong Kong property developers including Wharf, Sino Group, g Kong Land, South Asia Real Estate, and Newly Built Construction Property, etc. and famous mainland property developers including Zhonghai, Longfor, Country Garden, ShiMao and COFCO, China Resources Land, Greenland, China Railway Construction Real Estate Group, and OTC etc.

Its design projects and works can be seen all over New York, Hong Kong, Macao, Beijing, Shanghai, Guangzhou, Shenzhen, Xiamen, Tianjin, Chengdu, Chongqing, Shenyang and many other cities and areas.

图书在版编目（CIP）数据

设计有约 8 / 方峻 著 . – 武汉 : 华中科技大学出版社 , 2016.11

ISBN 978-7-5680-2265-1

Ⅰ . ①设… Ⅱ . ①方… Ⅲ . ①室内装饰设计 Ⅳ . ① TU238

中国版本图书馆 CIP 数据核字（2016）第 238346 号

设计有约 8
Sheji Youyue 8

方峻 著

出版发行：华中科技大学出版社（中国·武汉）　　　电话：（027）81321913
　　　　　武汉市东湖新技术开发区华工科技园　　　　邮编：430223

责任编辑：熊纯　　　　　特邀编辑：董莉婷　　　　排版设计：筑美文化
责任校对：赵营涛　　　　封面设计：王伟　　　　　责任监印：张贵君

印　　刷：中华商务联合印刷 (广东) 有限公司
开　　本：889 mm × 1194 mm　1/12
印　　张：37（上册 18 印张，下册 19 印张）
字　　数：222 千字
版　　次：2016 年 11 月第 1 版 第 1 次印刷
定　　价：558.00 元（USD 111.99）

投稿热线：13710226636　　　duanyy@hustp.com
本书若有印装质量问题，请向出版社营销中心调换
全国免费服务热线：400-6679-118 竭诚为您服务

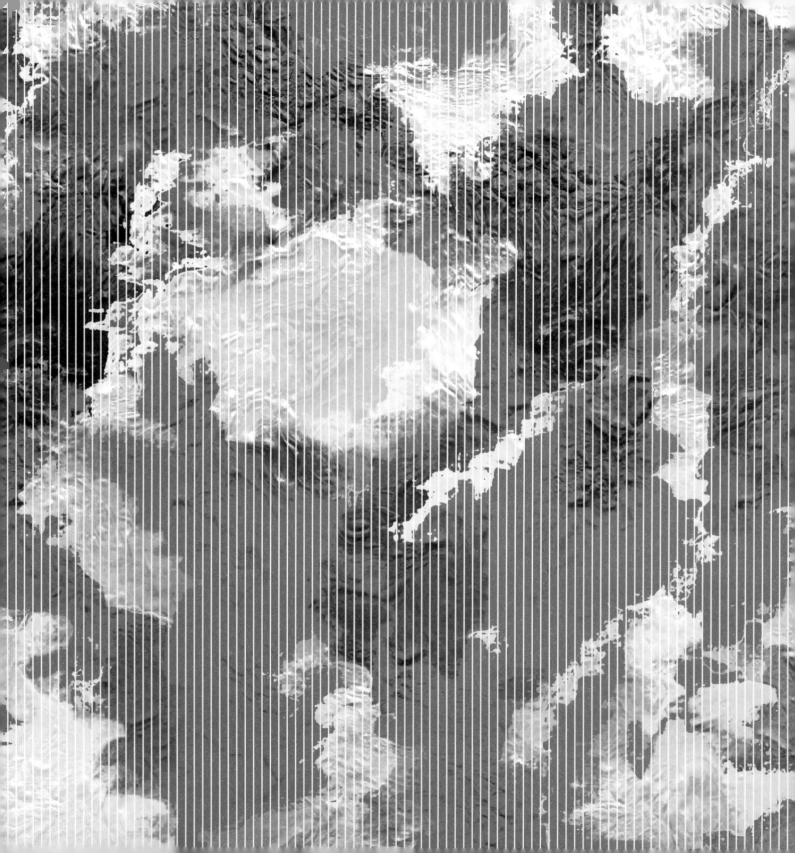